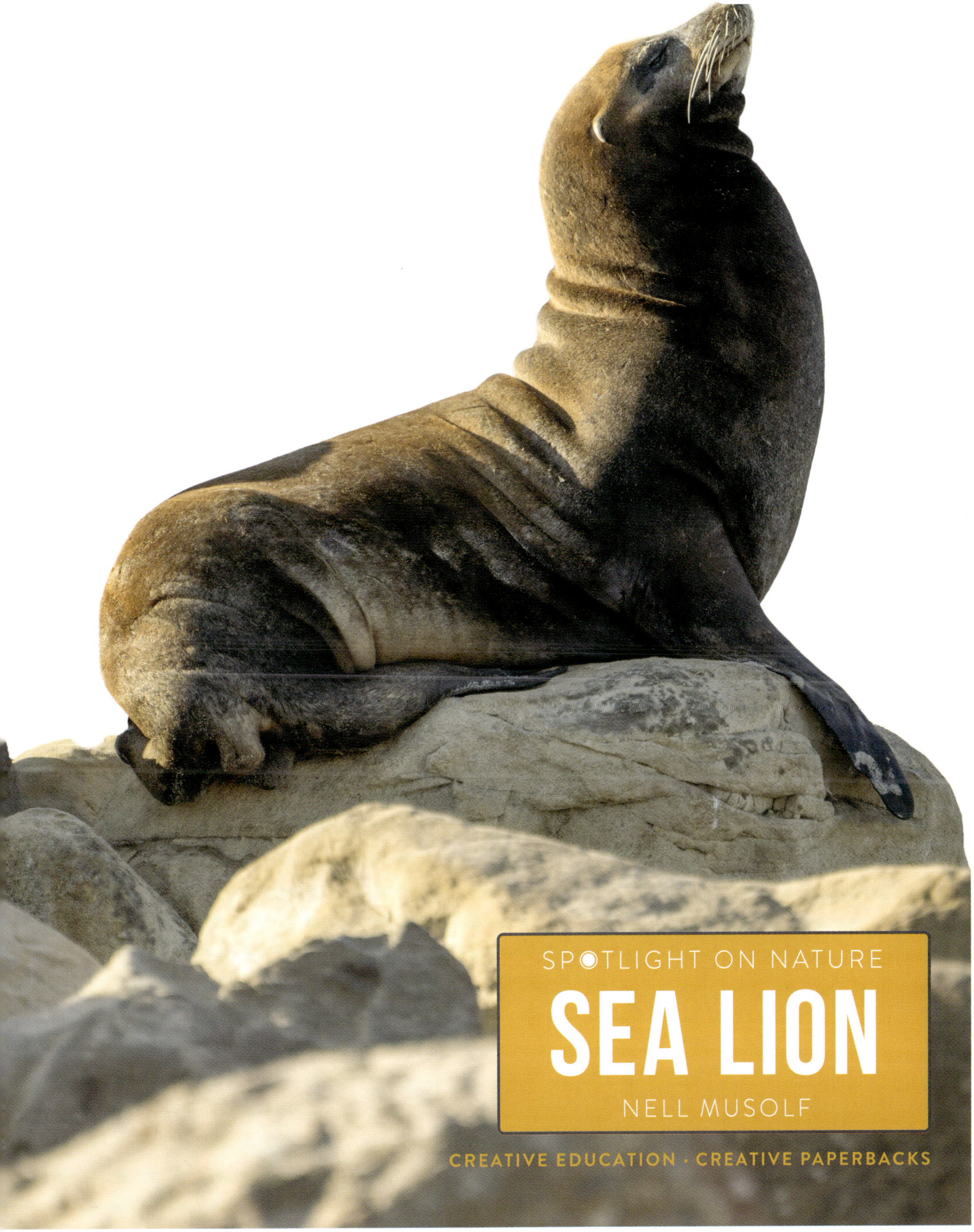

SPOTLIGHT ON NATURE

SEA LION

NELL MUSOLF

CREATIVE EDUCATION · CREATIVE PAPERBACKS

Published by Creative Education and Creative Paperbacks
P.O. Box 227, Mankato, Minnesota 56002
Creative Education and Creative Paperbacks
are imprints of The Creative Company
www.thecreativecompany.us

Design by Blue Design, Inc.
Art direction by Graham Morgan
Edited by Grace Beltowski

Photographs by Dreamstime/Christian Schmalhofer, 18, Derek Holzapfel, 29, Ivonne Wierink, 28; flickr/Biodiversity Heritage Library, 3, 6, 8, 10, 14, 16, 20, 22, 28; Getty Images/Doug Steakley, 4–5, Paul Souders, 11, Lynne Gilbert, 17; Pexels/Brian Hackworth, 27, Elianne Dipp, 14, Lachlan Ross, 9; Shutterstock/Anne M Vallone, 29, Katrina Outland, 10, rebvt, 20–21; Unsplash/Hc Digital, 24, Kace Rodriguez, 12, Mike Kitchen, cover, 1, Romy Vreeswijk, 16; Wikimedia Commons/Alexdi at English Wikipedia, 6, Andrew Turner, 21, Jennyhjert, 23, public domain, 29, Rhododendrites, 15

Every effort has been made to contact copyright holders for material reproduced in this book. Any omissions will be rectified in subsequent printings if notice is given to the publisher.

Copyright © 2026 Creative Education, Creative Paperbacks
International copyright reserved in all countries.
No part of this book may be reproduced in any form
without written permission from the publisher.

Library of Congress Cataloging-in-Publication Data
Names: Musolf, Nell, author.
Title: Sea lion / Nell Musolf.
Description: Mankato, Minnesota : Creative Education / Creative Paperbacks ; [2026] | Series: Spotlight on nature | Includes bibliographical references and index. | Audience: Ages 10-13 | Audience: Grades 4-6 | Summary: "Paddle into the sea lion's world with this nature title for middle-grade wildlife lovers. Informational text pairs with a narrative about a single sea lion family to spotlight the wild marine animal's life cycle, supported by infographics and a timeline of developmental milestones"— Provided by publisher.
Identifiers: LCCN 2024048419 (print) | LCCN 2024048420 (ebook) | ISBN 9798889896272 (library binding) | ISBN 9781682777930 (paperback) | ISBN 9798889897071 (ebook)
Subjects: LCSH: Sea lions—Juvenile literature.
Classification: LCC QL737.P63 M87 2026 (print) | LCC QL737.P63 (ebook) | DDC 599.79/75—dc23/eng/20241220
LC record available at https://lccn.loc.gov/2024048419
LC ebook record available at https://lccn.loc.gov/2024048420

Printed in India

CONTENTS

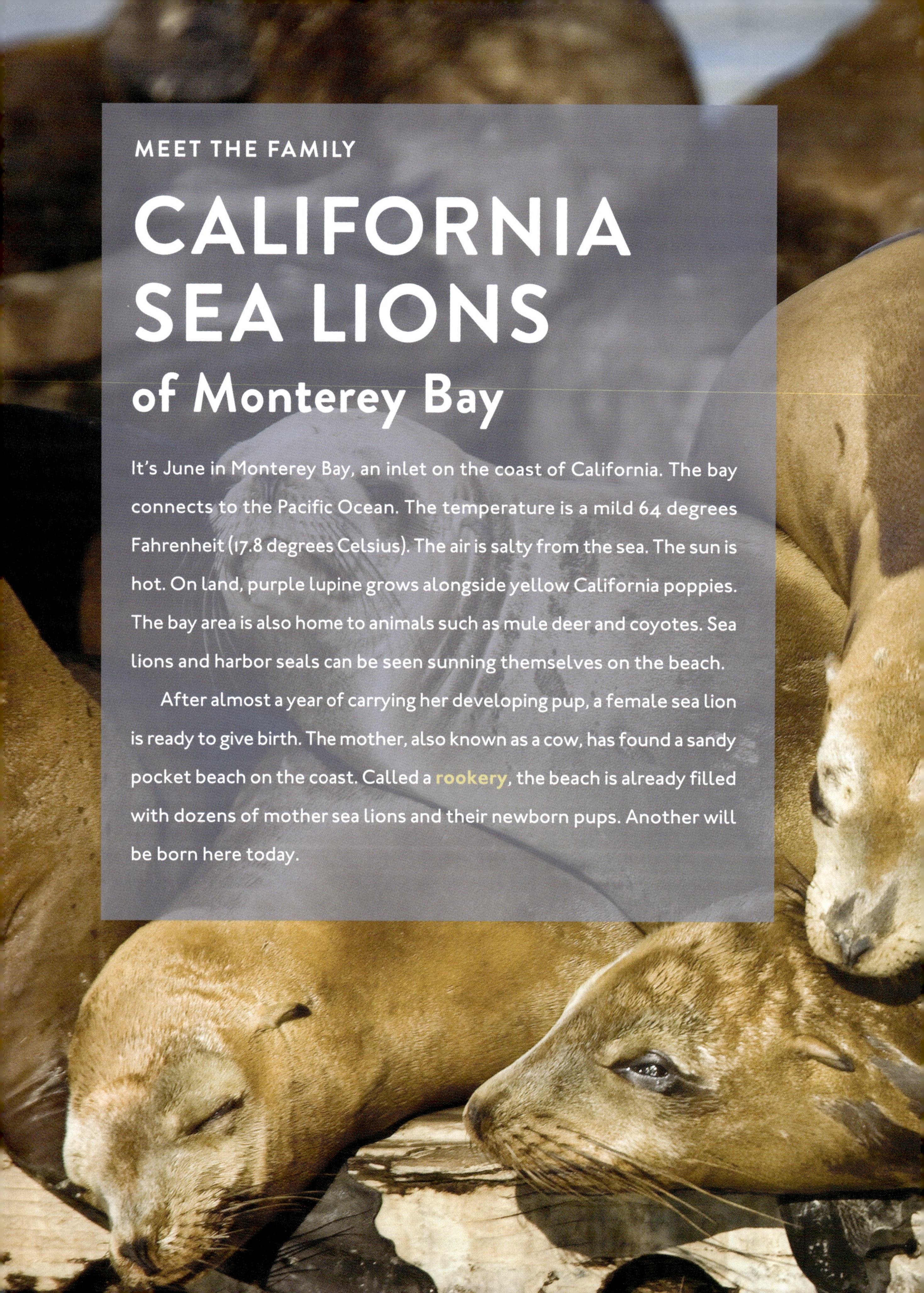

MEET THE FAMILY

CALIFORNIA SEA LIONS of Monterey Bay

It's June in Monterey Bay, an inlet on the coast of California. The bay connects to the Pacific Ocean. The temperature is a mild 64 degrees Fahrenheit (17.8 degrees Celsius). The air is salty from the sea. The sun is hot. On land, purple lupine grows alongside yellow California poppies. The bay area is also home to animals such as mule deer and coyotes. Sea lions and harbor seals can be seen sunning themselves on the beach.

After almost a year of carrying her developing pup, a female sea lion is ready to give birth. The mother, also known as a cow, has found a sandy pocket beach on the coast. Called a **rookery**, the beach is already filled with dozens of mother sea lions and their newborn pups. Another will be born here today.

CLOSE-UP

Fur baby

Sea lions' skin is covered in a layer of hair, or fur. The thickness of the hair varies in each species. When wet, a sea lion's fur feels like wet dog fur. It is soft and smooth.

CHAPTER ONE

LIFE BEGINS

Sea lions are **pinnipeds**. A pinniped is a marine mammal that has front and rear flippers. Other pinnipeds are seals and walruses. All pinnipeds live in the ocean, but they can also live on land for short periods of time. They rest on sandy beaches or rocky shores. They hunt for prey and cool off in the water.

There are six species of sea lions found around the world: Australian, California, Galapagos, New Zealand, South American, and Steller sea lions. Although sea lions **migrate**, they are named according to their primary locations. The Steller sea lion is the only exception. It is found in the North Pacific Ocean from Japan to Alaska, but it can also be found as far south as California. There are slight variations among the species. South American sea lions have shorter and wider muzzles than other species. Australian sea lions have lighter manes. Steller sea lions can be a reddish-brown color, while other species are dark brown or black.

CALIFORNIA SEA LION MILESTONES

DAY 1

- Born
- Weight: 16 pounds (7.3 kilograms)
- Length: 30 in (76.2 centimeters)
- Eyes are open
- Vocalizes to mother
- Can walk by the end of the day

All sea lions have external ear flaps, long flippers, and large chests and bellies. Their fur is short and thick. A layer of blubber keeps them warm in the ocean. Adult female sea lions weigh between 200 and 400 pounds (91–181 kg). Adult males weigh between 600 and 800 pounds (272–363 kg), but some weigh as much as 1,000 pounds (454 kg). Steller sea lions, which are the largest, can weigh up to 2,500 pounds (1,134 kg). They can reach lengths of 11 feet (3.4 meters).

CLOSE-UP

Special bonding

Mother sea lions can find their pups even among hundreds of other sea lions. This is because of the pup's unique bark, which the mother has known since the pup was born. After the mother locates her pup, she sniffs it to guarantee she has the right one.

Sea lions are important to ocean ecosystems. They eat a wide variety of fish, which helps regulate fish populations. California sea lions hunt for sardines, anchovies, octopus, and squid. Australian sea lions eat blue-throated

Welcome to the World

Birth is fast for sea lions. It can take a few minutes to an hour. As the cow feels her pup emerging from her body, she begins vocalizing to it. Her special sounds are answered by the pup. They talk to each other for 20 minutes. This talking helps the cow and pup bond. The cow smells, nuzzles, pulls, and nips her newborn. The pup is a dark shade of brown. Her eyes are open, and she studies her mother's face. The cow and pup are getting to know each other.

wrasses, squid, cuttlefish, and fairy penguins. South American sea lions eat octopus, squid, and **crustaceans**. Sea lions are also a food source for killer whales and various types of sharks, including great white, hammerhead, and blue sharks. Occasionally, one species of sea lions will prey on another, but this is rare.

Since sea lions live both in the water and on land, they play another important role as a nutritional **courier**. When a sea lion leaves the ocean, it carries nutrients from the water to the land. These nutrients keep the soil healthy.

3 WEEKS

- Begins interacting with other pups
- Has first swimming lesson

2 MONTHS

- **Molts** for the first time

CLOSE-UP

Flinging flippers

Unlike the rest of a sea lion's body, which has a thick layer of blubber to keep it warm, flippers aren't well **insulated**. A sea lion can regulate its body temperature by lifting a flipper out of the water to either absorb or release heat.

FEATURED  FAMILY

First Meal

The newborn pup is hungry. The mother nurses the pup, something she will do for the next 6 to 12 months. She won't leave her pup this first day, or the next. On the third day, she might leave the pup alone while she **forages** in the sea. Foraging helps the mother get her energy back. When she returns to the pup, she will nurse her again. Milk will be the only food the pup gets for a few months. When the pup is old enough, fish will be added to the menu.

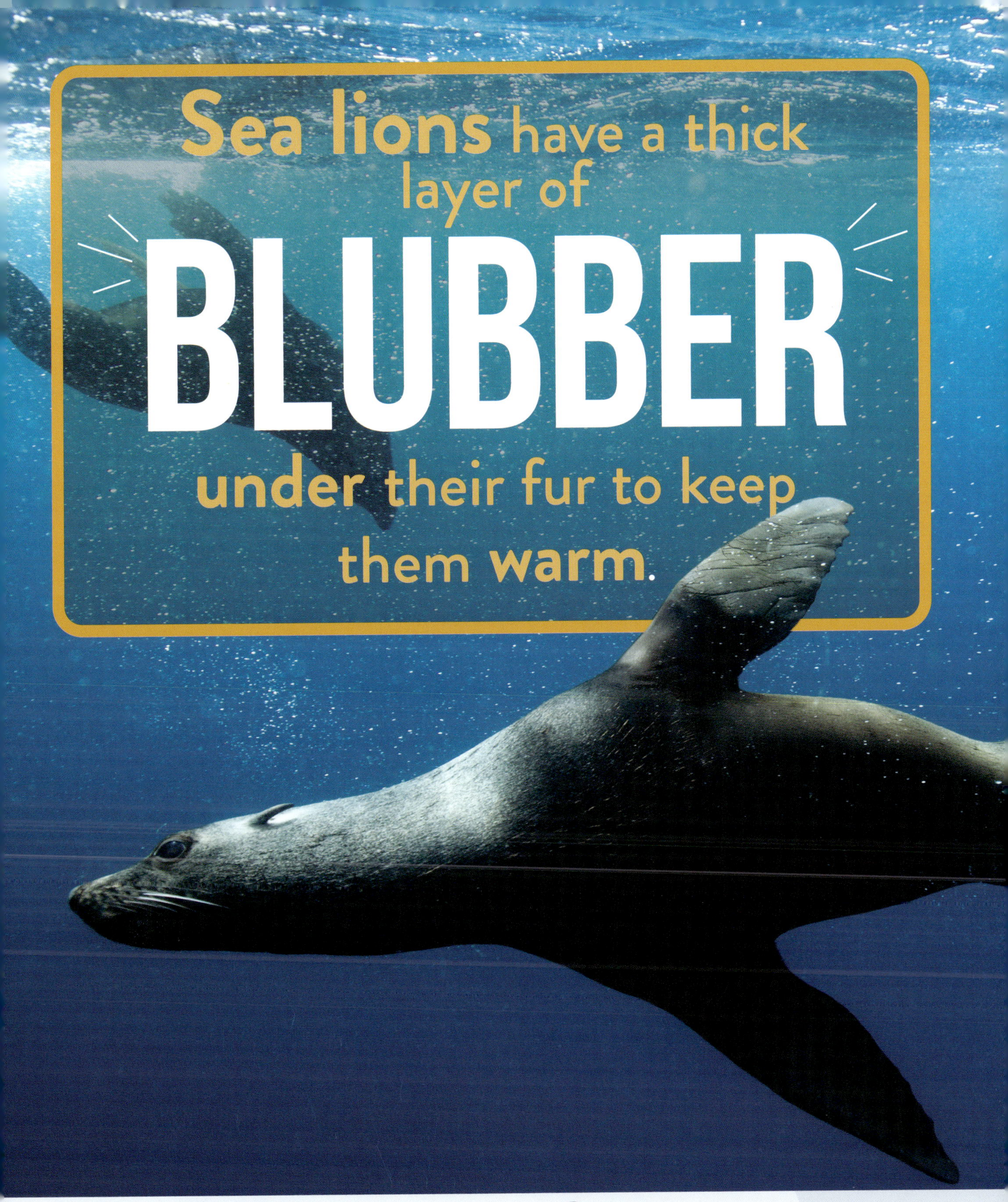

Sea lions have a thick layer of

BLUBBER

under their fur to keep them **warm**.

③ **MONTHS**

- Continues to nurse
- Starts to eat fish

⑥ **MONTHS**

- Molts a second time

CLOSE-UP

Molting

All sea lions molt. During this time, they shed their coats and grow new skin and hair. Molting protects the sea lions' skin from damage caused by fights and too much sun. Sea lions stay on land during this process.

CHAPTER TWO

EARLY ADVENTURES

Mother sea lions remain with their young pups for the first few days of life. Then, the mother temporarily leaves the pup on land while she goes back into the ocean. She spends up to five days feeding and foraging. While she is gone, the pup doesn't eat. The mother eventually returns to nurse the pup again. This pattern continues for up to a year.

How long a pup nurses depends upon its species. Australian sea lions nurse for up to 15 months and form strong bonds with their mothers during that time. Galapagos sea lions might nurse for up to three years. California sea lions are fully **weaned** at one year old. Their diet then consists of fish and other seafood like octopus.

When pups reach two to three weeks, they start to interact with other pups in the rookery. Rookeries consist of multiple families. Pups learn many social skills from these groups. They chase each other, surf together,

(1) YEAR

- Fully weaned from mother
- Starts hunting for its own prey

CLOSE-UP

Clear view

It is believed that sea lions don't see in color. Instead, they see in a spectrum of blue to green. Sea lions have a membrane called a tapetum lucidum in the back of their eyes. This membrane allows them to see more clearly in low light and at night.

FEATURED  FAMILY

Look Who's Walking

Since she was born, the pup has been using her four flippers to move through the water with ease. Her large front flippers enable her to move on land as well. After a swim, she hauls her body onto the shore. With some effort, she puts all four of her flippers under her body and begins walking. She's a little jerky, but she'll learn to move more smoothly with time. For now, she finds a warm rock on the beach and rests under the sun.

and sometimes push each other off rocks into the water. This kind of play prepares them for future territorial battles.

In the first six months of their lives, sea lions will molt two times. After that, they will molt yearly. As sea lions grow up, they become more independent. At around one year old, they usually no longer live with their mothers and may rarely interact with them. Male sea lions of the Galapagos species appear to remain closer to their mothers than females, possibly as protection.

(2) YEARS

- Diet is more varied, including fish, octopus, squid, clams, and lobster
- Length: 6.5 feet (2 m)

CLOSE-UP

Whiskers

Sea lions have long, stiff whiskers. They measure up to 11 inches (27.9 cm) in length. These whiskers have nerve fibers that make them sensitive to the surrounding ocean environment. Whiskers help sea lions navigate in the water and find prey.

FEATURED FAMILY

Give It a Try!

Hey, wait up! The six-month-old pup joins a group of older sea lions who are looking for a meal. Together, they go out into the sea to hunt for fish. As a group, they herd a family of sardines into a ball. The hungry pup has learned to hunt by observing the other sea lions. She watches them eagerly as they take turns grabbing a fish to eat. Finally, it's her turn for a fish. She lunges forward and snatches one quickly. Just in time because she is starving!

4 YEARS

- Females reach sexual maturity and breed for the first time
- Female weight: 300 pounds (136 kg)

CLOSE-UP

Ear plugs

Sea lions have external ear flaps. These flaps act like ear plugs. They turn downward. This prevents water from getting inside their ears. All six species of sea lions have external ear flaps.

CHAPTER THREE

LIFE LESSONS

After leaving the rookery, sea lions continue to be social. They are often found in groups on the water called herds or **rafts**. Rafts can have over 1,000 members. The members stay close together, sometimes lying on top of each other when they are sleeping. In deep water, sea lions may also sleep by floating with their noses out of the water. This allows them to breathe. In shallow areas, they sink to the bottom while sleeping. They rise to the surface when they need air.

Males and female sea lions live together in the rafts. Sea lions start breeding around age four or five. When mating season begins in late June, the males, or bulls, are the first out of the water. They need extra time to prepare. First, they stake out a territory. The territory might be on ice, rocks, or sand. Once a bull has established his own breeding territory, he competes with other males by barking loudly and aggressively.

5 YEARS

- Males reach sexual maturity and breed with several females
- Male weight: 700 pounds (318 kg)
- Females have second pup

Male sea lions bark continuously during the breeding season. This barking helps them defend their territories and intimidate other males in the area. They might also stare, shake their heads, or lunge at other males to keep them away. Male sea lions don't eat during mating season.

Female sea lions usually breed once per year. Male sea lions find as many females as possible. A single

FEATURED FAMILY

This Is How It's Done

The year-old sea lion is hungry. She has learned how to hunt for food from her mother and other sea lions. Now she is hunting on her own. She moves smoothly through the cold ocean water. What's that? She spots a school of herring swimming near her. Her mouth opens as she moves forward, collecting water on the way. Snap! She closes her mouth quickly. Gulp! She has plenty of teeth, but she swallows her prey whole. Satisfied with her snack, she moves on. What will she find next?

CLOSE-UP

Scratch that itch!

Sea lions have three toenails on their hind flippers that they use for scratching. Their flexibility allows their flippers to reach almost all over their bodies. If they can't reach an itchy spot, they rub against rocks or other sea lions.

10 YEARS

- Continues to molt each year
- Females give birth to one pup each year
- Length: Up to 7.5 feet (2.3 m)

bull mates with an average of 16 cows every mating season. He forms a **harem**. The male will watch over his harem and keep the female sea lions safe. Often, several harems exist in the same area. After the breeding season ends in early August, females typically stay together. Male adults and juveniles leave the rookery and migrate to other feeding sites for the winter.

CLOSE-UP

Layers of teeth

Every year, a sea lion develops a new layer of teeth. It's possible to tell how old a sea lion is by counting the layers. This is like counting the rings in a tree trunk.

Most sea lions live to around 20 years in the wild. They have been known to live up to 30 years in captivity. Sea lion pups have a **mortality rate** of 10 to 15 percent during the first month of life. Some newborns are abandoned and die. Some are swept away from rookeries by rough ocean waves. Others become ill and die before reaching one year of age.

Practice Makes Perfect

Fully grown, the sea lion is now a master at hunting on her own. She knows where to find prey and how to catch it. She moves through the water, her flippers helping her keep a steady pace. She spots a flatfish and swims behind it. With one smooth move, the flatfish is in her mouth. Then she emerges from the water onto the shore. She begins shaking her head back and forth. Pieces of the flatfish break off. Soon it is small enough for her to swallow. Delicious!

20 YEARS

- End of life

CHAPTER FOUR

HELPING SEA LIONS SURVIVE

For hundreds of years, sea lions were hunted by humans around the world. Along the coast of Alaska, Steller sea lions were hunted for their meat, oil, blubber, and hides. The hides were used to cover boats that were called skin boats. Sea lions were also killed for **bounties**. People thought there were too many of them. In 1972, the Marine Mammal Protection Act was passed. It became illegal to hunt, capture, and kill marine mammals, including sea lions, in the United States. Sea lions are still in danger. They can get caught in drift nets that are put in the ocean by fisheries. These drift nets catch fish for commercial businesses.

Another threat to sea lions is climate change. As water temperatures rise due to global warming, currents in the water slow down and less fish are available for sea lions to eat. With their food sources threatened, it could become harder for sea lions to survive. Certain forms of algae also contain a toxin that attacks the brain and heart of sea lions. The algae are eaten by small fish that are later eaten by sea lions. The toxin accumulates in the sea lion's body, eventually causing health issues.

There are ways people can help sea lions. One solution is not using chemicals on grass or gardens. These chemicals end up in rivers, lakes, and oceans where they are ingested by animals. Another way is to properly dispose of plastic. Plastic bags and rings found on six-packs of soda often end up in oceans and can be eaten by sea lions or wrapped around them. Avoiding plastic or cutting it up before throwing it away are good ways to keep marine animals safe.

People can also volunteer with organizations dedicated to protecting ocean life. California's Marine Mammal Center has over 1,000 volunteers who work year round to help sea lions stay healthy and alive. Youth volunteers, age 15 to 18, are trained to help by washing equipment and cleaning animal pens and pools. They learn how to talk to others about ocean conservation.

These actions help sea lions and all ocean animals. They also help the planet. Starting with even a small step is a step in the right direction.

FAMILY ALBUM

SNAPSHOTS

Galapagos sea lions are the smallest sea lion species. Males weigh up to 550 pounds (250 kg) and females weigh between 110 and 220 pounds (50–100 kg).

Male Galapagos sea lions have a large bump on their foreheads, making them easy to spot.

South American sea lions grow the largest mane of all sea lion species. This mane makes them look more like lions.

Australian sea lions are the rarest species of sea lion. There are estimated to be 10,000 to 12,000 in the wild.

Female Australian sea lions have the longest pregnancy at 17 and a half months.

There was a seventh species of sea lion called the **Japanese sea lion**. This species became extinct in the 1970s. There have been a few reported sightings since then, but none have been verified.

Sea lions sometimes eat rocks during fasting periods. Scientists believe that is because the rocks give a feeling of fullness when sea lions can't eat fish or other prey.

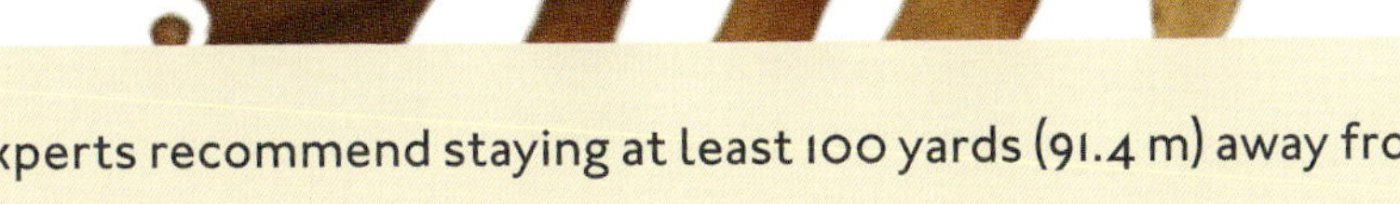

Experts recommend staying at least 100 yards (91.4 m) away from sea lions to give them space. If a sea lion is staring at a human, it shows that it feels threatened.

Sea lions can move as quickly as 35 miles (56.3 km) per hour in the water.

Sea lions are found in oceans around the world except for the North Atlantic. The water temperatures and available prey are ideal for sea lions, but scientists have yet to uncover why they avoid that area of ocean water.

WORDS to Know

bounty money given for capturing or killing an animal

courier a person or animal who carries something from one place to another

crustacean a water creature, such as a lobster or crab, that has a hard outer shell

forage search for food

harem a group of female animals sharing a single mate

insulated traps and retains heat

migrate move from one place to another during the year, usually for feeding or breeding

molt shed hair, skin, feathers, or an outer layer periodically

mortality rate the death rate of individuals in a population over a specific period of time

pinniped a carnivorous aquatic mammal with four flippers for limbs

raft a congregation of animals resting on water

rookery a breeding ground for mammals, such as sea lions

wean accustom an infant to food other than its mother's milk

LEARN MORE

Books

Adamson, Heather. *Sea Lions*. Minneapolis: Bellwether Media, 2018.

Gish, Melissa. *Sea Lions*. Mankato, Minn.: Creative Education and Creative Paperbacks, 2024.

Rustad, Martha E. H. *All About Baby Sea Lions*. North Mankato, Minn.: Pebble, 2022.

Websites

"California Sea Lion." National Geographic Kids.

https://kids.nationalgeographic.com/animals/mammals/facts/california-sea-lion

"Sea Lion." Britannica Kids.

https://kids.britannica.com/kids/article/sea-lion/353752

"Sea Lion Facts for Kids." Dolphin Research Center.

https://dolphins.org/kids_sea_lion_facts

Documentaries

Chidanand, Aadya, dir. *A Second Chance*. Los Angeles: USC Impact, 2023.

McCarten, Amelia, and Paul Phelan. *Sea Lions: Life by a Whisker*. Moore Park, Aus.: Definitions Films, 2020.

Smith-Baker, Nick, Melanie Gerry, and James Hemming, dirs. *Growing Up Animal*. Season 1, episode 2, "A Baby Sea Lion's Story." Burbank, Calif.: Disney Plus, 2021.

Note: Every effort has been made to ensure that any websites listed above were active at the time of publication. However, because of the nature of the Internet, it is impossible to guarantee that these sites will remain active indefinitely or that their contents will not be altered.

Visit

AUDUBON ZOO

Observe sea lions swimming and playing from the underwater viewing area.

6500 Magazine Street
New Orleans, LA 70118

COLUMBUS ZOO AND AQUARIUM

Take a tour of the aquarium's seal and sea lion facility to learn how the zoo cares for marine animals.

4850 West Powell Road
Powell, OH 43065

MINNESOTA ZOO

Check out the California sea lions at the Discovery Bay exhibit.

13000 Zoo Boulevard
Apple Valley, MN 55124

POINT REYES PENINSULA

Travel to the Sea Lion Overlook for a glimpse of sea lions in the wild.

West Sir Francis Drake Boulevard
Inverness, CA 94937

INDEX